Graphic Organizers in Science™

Learning About Simple Machines with Graphic Organizers

Jonathan Kravetz

The Rosen Publishing Group's
PowerKids Press™
New York

For Andrew, who loves machines

Published in 2007 by The Rosen Publishing Group, Inc.
29 East 21st Street, New York, NY 10010

First Edition

Editor: Jennifer Way
Layout Design: Julio A. Gil

Photo Credits: Cover & title page (center) © Ole Graf/zefa/Corbis; cover & title page (top left), pp. 7 (top right), 11 (center) Cindy Reiman; cover & title page (top right), p. 7 (bottom right) © www.istockphoto.com/Nathan Watkins; cover & title page (bottom left), p. 7 (bottom center left) © www.istockphoto.com/Robert Lerich; cover & title page (bottom right), p. 7 (bottom center right) © www.istockphoto.com/Kerstin Klaassen; pp. 7 (top left), 20 (left) © www.istockphoto.com/Tom Mc Nemar; pp. 7 (top center), 20 (right) © www.istockphoto.com/blaneyphoto; p. 7 (bottom left) © www.istockphoto.com/Abel Filho; p. 8 © Index Stock/Corbis; p. 11 (left) © Steve Smith/Getty Images; p.11(right) © Anne Ackermann/Getty Images; p. 12 © Ed Bock/Corbis; p. 19 (left) © H. Schmid/zefa/Corbis; p. 19 (right) © David Stoecklein/Corbis; p. 20 (center left) © www.istockphoto.com/Réne Mansi; p. 20 (center right) Maura B. McConnell.

Library of Congress Cataloging-in-Publication Data

Kravetz, Jonathan.
  Learning about simple machines with graphic organizers / Jonathan Kravetz.— 1st ed.
      p. cm. — (Graphic organizers in science)
  Includes index.
  ISBN 1-4042-3411-X (library binding) — ISBN 1-4042-2206-5 (pbk.) — ISBN 1-4042-2396-7 (six pack)
  1. Simple machines—Juvenile literature. I. Title. II. Series.

TJ147.K73 2007
621.8—dc22

                                    2005032938

Manufactured in the United States of America

Contents

KWL Chart: Simple Machines

What I **K**now	What I **W**ant to Know	What I've **L**earned
• There are many different kinds of machines.	• What makes something a simple machine?	• Simple machines need only one force to work.
• People do work.	• Why do people use simple machines?	• Simple machines make doing work easier.
• Simple machines help people do many tasks.	• What happens if a simple machine cannot do the job?	• Simple machines can combine with other simple machines to make complex machines.

What Are Simple Machines?

A machine is a tool used to make work easier. Simple machines are simple because they need a single **force** to work. When you use a simple machine to move an object, you are actually doing the same amount of work you would use to move the object without the machine. However the machine makes work easier by taking the amount of force that you apply to the object and making that force greater. An **inclined plane**, such as a freeway on-ramp, makes it easier to move your car up onto the freeway. A shovel, which is a type of **lever**, makes it easier to lift a pile of dirt. Simple machines can form **complex** machines when combined with other simple machines. You will learn more about this in chapter 10.

This graphic organizer is called a KWL chart. Making a KWL chart can help you find out what you already know, what you want to know, and what you learn from studying a subject. This KWL chart shows some things you can learn about simple machines.

Types of Simple Machines

This book will teach you about seven basic types of simple machines. They are the lever, the wheel and **axle**, the **pulley**, the gear, the inclined plane, the **wedge**, and the screw. The wedge and the screw are really types of inclined planes, but they look different from inclined planes and so are considered to be different simple machines.

People use simple machines every day. The doorstop that holds your door in place is a wedge. The moving parts in your grandfather clock are gears. The steering wheel in your car uses a wheel and axle. A seesaw is a lever. The steps you use to walk up or down are a type of inclined plane. In the following chapters, we will see how all these simple machines work.

This graphic organizer is called a concept web. Concept webs are used to organize facts about a subject. The subject goes in the middle, and the facts go around it. Here the subject is simple machines.

Concept Web: Types of Simple Machines

Screw: A screw is really an inclined plane wrapped around a nail. A drill bit is a type of screw.

Wedge: A wedge is two inclined planes joined back to back. An ax uses a wedge.

Pulley: A pulley is made up of a rope or chain wrapped around a wheel. Cranes use pulleys.

Lever: A lever is a rod that turns at a fixed point. The fixed point is called a fulcrum. A shovel is a type of lever.

Simple Machines

Inclined Plane: An inclined plane is a simple machine with a slanted surface. A ramp is an inclined plane.

Wheel and Axle: A wheel and axle is made of a large wheel connected to a post or axle. A wheelbarrow uses a wheel and axle.

Gear: Gears are wheels with sharp teeth around the edge. Gears are used in different types of machines, such as clocks and bicycles.

Sequence Chart: How a Screwdriver Works

1. You turn the handle. Your effort is a force.

2. The thin metal rod at the other end of the screwdriver turns. The effort of turning the handle creates greater force on the metal rod.

3. The screw turns. The greater force created by turning the handle makes it easier to overcome the load applied by the screw.

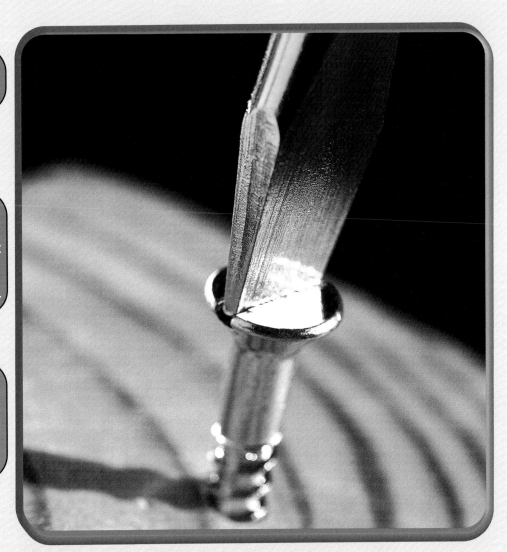

How Simple Machines Work

To move an object, you need to overcome a force called the load. A simple machine helps you overcome the load of an object. It takes your effort and helps you use it in a less wasteful way. The force you apply is your **effort**. If you moved a load of bricks on your own, your back, legs, and arms would do most of the job. Using a simple machine, such as a wheelbarrow, would make moving the bricks easier.

The **ratio** of load to effort is known as the **mechanical advantage**. This ratio is defined as the load **divided** by the amount of effort used. If a machine has a four-to-one force ratio, that means the machine will overcome a force four times greater than the effort you use.

This graphic organizer is called a sequence chart. Sequence charts show the steps of something in order. This chart shows how the force applied to the handle of a screwdriver creates a greater force at the screwdriver's head, making it easier to turn the screw.

Levers

A lever is a rod that moves around a fixed point. The fixed point is called a **fulcrum**. In order to move an object using a lever, you must apply some force to the lever. Some basic tools use levers, including **scissors**, wheelbarrows, and pliers.

There are three types of levers. Class-one levers have the fulcrum between the effort and the load. Shovels and seesaws are class-one levers. Class-two levers have the load between the effort and the fulcrum. Wheelbarrows and car doors are class-two levers. Class-three levers have the effort between the load and the fulcrum. Hockey sticks and tweezers are class-three levers. The closer your fulcrum is to the load, the more force the lever produces. For example, the closer to the end of the handle you hold a **wrench**, the easier it is to turn a bolt.

A compare/contrast chart is a graphic organizer that helps you see the ways in which things are alike and different. This compare/contrast chart compares different types of levers.

Compare/Contrast Chart: Levers

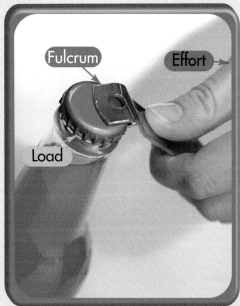

Simple Machine	Number of Levers Used	Class of Levers Used
Seesaw	1	Class 1
Hammer's Claw	1	Class 1
Scissors	2	Class 1
Pliers	2	Class 1
Wheelbarrow	1	Class 2
Nutcracker	2	Class 2
Fishing Rod	1	Class 3
Tweezers	2	Class 3
Tongs	2	Class 3

Cause-and-Effect Chart: Wheel and Axles

Cause		Effect
The wagon rests on top of the axles.	→	A person applies force to the wheel by pulling the handle.
The wheel turns the axle.	→	The wagon moves.

Wheel and Axles

The wheel and axle is a simple machine made of a large wheel attached to a rod, or axle. A wheel and axle is really a type of lever that moves in a circle around a fulcrum. Wheel and axles have many uses in daily life. Cars, pencil sharpeners, fans, and in-line skates all use the wheel and axle.

There are two basic types of wheel and axles. In one type effort is applied to the wheel. Then the effort is turned into a bigger force by the axle. Steering wheels and screwdrivers work this way. In the second type of wheel and axle, effort is applied to the axle. Then the effort turns the wheel. The force applied to the axle turns the wheel farther than the axle because the wheel is bigger. Cars wheels and eggbeaters use this type of wheel and axle.

A cause-and-effect chart lists causes on the left and effects on the right. Causes are things that happen. Effects are things that happen as a result of a cause. This chart explains how a wagon works.

Pulleys

The pulley is a simple machine made of a rope or chain wrapped around a wheel. As the wheel turns, the rope or chain moves. Flagpoles and window blinds use pulleys.

Pulleys can be used to lift heavy loads. The load is connected to the rope and force is applied at the other end. When force is applied, the rope and wheel move, lifting the load. Pulleys allow you to use your body weight to help you lift. Multiple pulleys can be combined in what is called a block and tackle. That can make lifting even easier. Each pulley you add breaks the rope into another **section**. Each section holds some part of the weight of the object you are lifting. This makes the mechanical advantage greater. Look at the flow charts on the opposite page to see how this works.

A flow chart uses pictures to show how something works. In these flow charts, we see how one wheel and two wheel pulleys work. The arrows in each picture show the direction in which the force moves.

Flow Chart: Pulleys

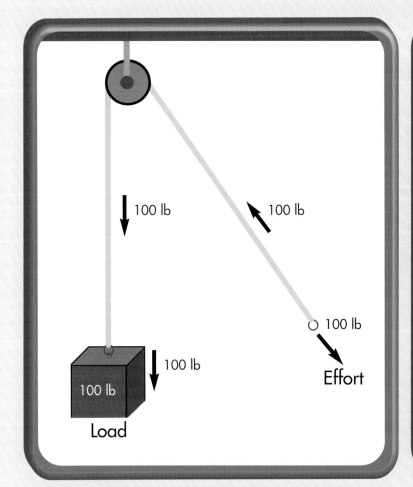

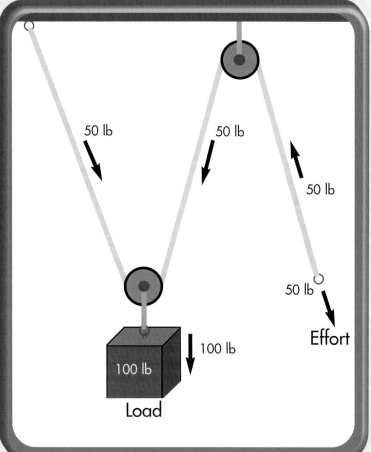

Lifting 100 pounds (45 kg) with one pulley means you must apply 100 pounds (45 kg) of force on the rope to lift it.

If you add a second pulley, then the weight is hanging by two sections of rope. Each section is now holding only 50 pounds (23 kg), so when you pull down you have to apply only 50 pounds (23 kg) of force to lift the object.

Diagram: How a Clock Works

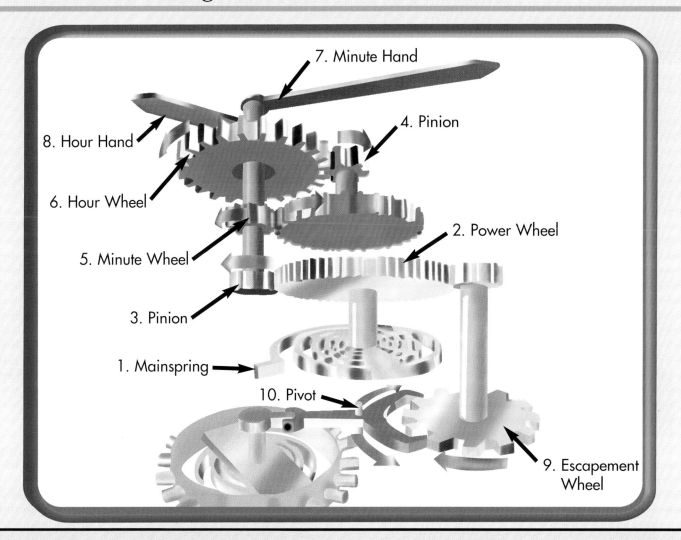

7. Minute Hand

4. Pinion

8. Hour Hand

6. Hour Wheel

5. Minute Wheel

2. Power Wheel

3. Pinion

1. Mainspring

10. Pivot

9. Escapement Wheel

Clocks are powered by a mainspring, which when wound provides energy to run the clock. The force from the mainspring turns the power wheel. It pushes motion through the pinion gears, which turn the minute wheel and the hour wheel. This makes the clock's hands move. The escapement wheel controls the motion of the power wheel. The escapement is controlled by the back-and-forth movement of the pivot. The pivot's motion also produces the ticktock sound of a clock.

Gears

A gear is a wheel with teeth around the edge that allows it to fit together with other gears. When one gear turns, it turns the other in the opposite direction. Gears are often used to produce a certain speed. For example, a bicycle's top gear makes the bike go fast, and the lowest gear makes it go slow. Gears do this by changing the amount of force. In any two gears, the larger one will turn more slowly than the smaller one but with more force.

Gears are everywhere there are engines producing a turning motion. Car engines have lots of gears in them. VCRs have many gears, as do many types of clocks.

A diagram is a drawing showing how the parts of something work. In this diagram we see how the gears inside a clock make it work. The numbers show the order in which the parts of a clock move.

Inclined Planes and Wedges

An inclined plane is a sloping surface used to lift an object. A ramp is an example of an inclined plane. When you move an object up an inclined plane, you move it a greater distance than you would if you lifted it straight up. Lifting the object this way means you use less force to do work. The **pyramids** in Egypt are made of huge stone blocks. For years people wondered how the Egyptians moved those blocks. Scientists now think the Egyptians pushed the blocks up inclined planes.

A wedge is two inclined planes joined back-to-back. Examples of wedges include a knife, an ax, and a **razor** blade. Wedges are usually used to cut or divide objects but can also be used to stop an object from moving. With enough **friction** a wedge, such as a doorstop, can keep an object from sliding.

A chart is a simple way to organize facts. This chart explains different kinds of inclined planes and wedges and the jobs they do.

Chart: Inclined Planes and Wedges

Ladder

Chisel

Object	Inclined Plane or Wedge	What It Does
Parking Ramp	Inclined plane	Used to move cars to upper levels in a parking lot
Ladder	Inclined plane	Used to climb
Doorstop	Wedge	Used to hold a door open
Chisel	Wedge	Used to break something off

Tree Chart: Inclined Planes

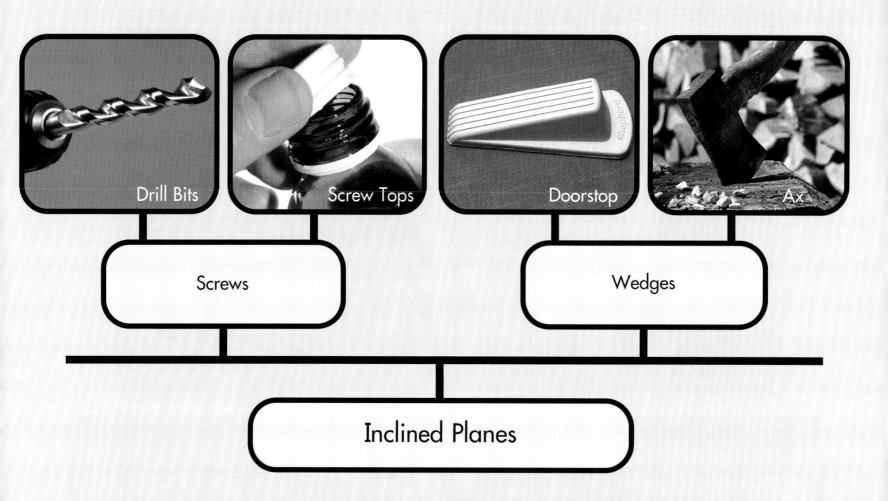

Drill Bits | Screw Tops | Doorstop | Ax

Screws | Wedges

Inclined Planes

Screws

A screw is an inclined plane wrapped around a cone. The **ridges** of the screw are called threads. Screws are often used to hold objects, such as pieces of wood, in place. The threads cut a **groove** in the wood as you turn the screw, making it hold very tightly. It also makes it hard to remove the screw by pulling it straight out. To remove a screw, you have to turn it in the opposite direction.

A turning screw changes the forces applied to it into a greater forward or backward force. That way the screw can easily drive into an object.

A tree chart shows the subject of the graphic organizer in the trunk. Elements of the subject are added as branches. This tree chart shows how other simple machines are like the inclined plane.

Machines and Us

It is believed that the lever was one of the first simple machines used as a tool by people. The first levers were probably branches that were used to lift things. Historians believe the ancient Mesopotamians used wheels as early as 3500 B.C. The Greek scientist Archimedes is believed to have created a screwlike tool for raising water from under the ground. Some also say Archimedes invented one of the first pulleys.

Today many everyday machines are really two or more simple machines working together. These are called complex machines. A wheelbarrow is a complex machine that uses a lever and a wheel and axle. Simple and complex machines continue to make work easier for people.

Glossary

axle (AK-suhl) A bar or a rod on which a wheel or a pair of wheels turns.

complex (kom-PLEKS) Made up of many connected parts.

divided (dih-VYD-ed) Broke apart or separated.

effort (EH-fert) The amount of force applied to an object.

force (FORS) Something that moves or pushes on something else.

friction (FRIK-shin) The rubbing of one thing against another.

fulcrum (FUL-krum) The point on which a lever turns.

groove (GROOV) A dent or cut in the surface of something.

inclined plane (in-KLYND PLAYN) A simple machine with a sloped surface.

lever (LEH-vur) A rod that turns at a fixed point.

mechanical advantage (meh-KA-nih-kul ud-VAN-tij) The ratio, or comparison, of load to effort in a simple machine.

pulley (PU-lee) A type of simple machine made of a rope or chain wrapped around a wheel.

pyramids (PEER-uh-midz) Large, stone structures with a square bottom and triangular sides that meet at a point on top.

ratio (RAY-shoh) A comparison between two things.

razor (RAY-zer) A sharp tool used for cutting hair.

ridges (RIJ-ez) The long, narrow, upper parts of things.

scissors (SIH-zerz) A cutting tool that has two blades.

section (SEK-shun) A part of something.

wedge (WEJ) Something shaped like a triangle.

wrench (RENCH) A tool used to turn objects, such as nuts.

Index

A
axle(s), 6, 13

D
doorstop, 6, 18

E
effort, 9–10, 13

F
force(s), 5, 9–10, 13–14,
   17–18, 21
friction, 18
fulcrum, 10, 13

G
gear(s), 6, 17

I
inclined plane(s), 5–6, 18,
   21

L
lever(s), 5–6, 10, 13, 22
load(s), 9–10, 14

M
mechanical advantage, 9,
   14
Mesopotamians, 22

P
pulley(s), 6, 14, 22
pyramids, Egyptian, 18

R
ratio, 9

S
screw(s), 6, 21
seesaw(s), 6, 10

W
wedge(s), 6, 18
wheel and axle(s), 6, 13,
   22

Web Sites

Due to the changing nature of Internet links, PowerKids Press has developed an online list of Web sites related to the subject of this book. This site is updated regularly. Please use this link to access the list:
www.powerkidslinks.com/gosci/simpmac/